Structures of Life

by Emily Sohn and Judy Kentor Schmauss

Chicago, Illinois

Norwood House Press
Chicago, Illinois
For information regarding Norwood House Press, please visit our website at www.norwoodhousepress.com or call 866-565-2900.

Contributors: Edward Rock, Project Content Consultant
Editor: Lauren Dupuis-Perez
Designer: Sara Radka
Fact Checker: Sam Rhodes

Photo Credits in this revised edition include: Getty Images: Graham Custance Photography, 12, Image Source, 15, iStockphoto, 4, Jose Luis Pelaez Inc, 7, Spencer Platt, cover, 1, stevanovicigor, 21, stock_colors, 23; Newscom: Dorling Kindersley, 25; Pixabay: Byunilho, background (paper texture), GDJ, background (tech pattern); Shutterstock: Jeff Whyte, 19, Monkey Business Images, 8, Robert Eastman, 18; Wikimedia: DinoTeam, 20

Library of Congress Cataloging-in-Publication Data
Names: Sohn, Emily, author. | Schmauss, Judy Kentor, author. | Sohn, Emily. iScience.
Title: Structures of life / by Emily Sohn and Judy Kentor Schmauss.
Description: [2019 edition]. | Chicago, Illinois : Norwood House Press, [2019] | Series: iScience | Audience: Ages 8-10. | Includes bibliographical references and index.
Identifiers: LCCN 2018057947 | ISBN 9781684509560 (hardcover) | ISBN 9781684043859 (pbk.) | ISBN 9781684043965 (ebook)
Subjects: LCSH: Fossils—Juvenile literature. | Paleontology—Juvenile literature. | Animals, Fossil—Juvenile literature. | Habitat (Ecology)—Juvenile literature.
Classification: LCC QE714.5 .S635 2019 | DDC 560—dc23
LC record available at https://lccn.loc.gov/2018057947

Hardcover ISBN: 978-1-68450-956-0
Paperback ISBN: 978-1-68404-385-9

Revised and updated edition ©2020 by Norwood House Press. All rights reserved.
No part of this book may be reproduced without written permission from the publisher.
320N—072019
Manufactured in the United States of America in Stevens Point, Wisconsin.

Contents

iScience Puzzle 6
Discover Activity 7
Where Did It Live? 9
What Did It Eat? 12
How Did It Stay Safe? 16
Connecting to History 19
Science at Work 21
The New Generation 22
Solve the iScience Puzzle 28
Beyond the Puzzle 29
Glossary 30
Further Reading/Additional Notes 31
Index 32

Note to Caregivers:
In this updated and revised iScience series, each book poses many questions to the reader. Some are open ended and ask what the reader thinks. Discuss these questions with your child and guide him or her in thinking through the possible answers and outcomes. There are also questions posed which have a specific answer. Encourage your child to read through the text to determine the correct answer. Most importantly, throughout the book, encourage answers using critical thinking skills and imagination. In the back of the book you will find answers to these questions, along with additional resources to help support you as you share the book with your child.

Words that are **bolded** are defined in the glossary in the back of the book.

Life in Rocks

Life comes in many shapes and sizes. Can you tell a plant from an animal? Sometimes, telling one from the other is harder than it sounds. That's especially true when you're looking at a **fossil.** Fossils are rocks. They show signs of once-living things. Bones and parts of plants can become fossils. First, they have to get buried under dirt or water. Over time, minerals move into spaces in the bones. This process turns the bones into remains that can last a long time. **Casts** of footprints and empty casts of sea creatures are fossils, too. Fossils can reveal clues about the past.

iScience Puzzle

Guess That Fossil

Your phone rings. A **paleontologist** is on the line. She's a scientist who studies plants and animals that lived in the past. "You've won a science contest!" she says.

The paleontologist has found a new fossil. She wants a student to help figure out what **organism** the fossil shows. And she is hosting a TV show about the discovery. You will be the guest star! Read on for clues about the fossil.

A paleontologist examines a fossil she found.

Discover Activity

Twenty Questions

Scientists spend a lot of time trying to solve mysteries. They begin by asking simple questions. Then they work to figure out the answers. Each answer can raise a whole bunch of other questions. Over time, the answers to the simple questions can help answer bigger questions. In this way, scientists learn new things about the world.

When you ask a question, you are part of the scientific process.

This activity will help you start thinking like a scientist. First, find a partner. Tell your partner to think of a person, place, or thing. Then ask questions that will help you guess what your partner is thinking of. Your partner can answer only with "Yes" or "No."

The goal is to find out what your partner is thinking by asking as few questions as possible. So your questions need to be specific. For example, it won't help much to ask, "Is it small?" You'll get a better answer from, "Is it smaller than my hand?"

Did you guess the right answer? How many questions did you need? Which questions worked best?

In this book, you will learn to ask good questions. The answers will help you solve the iScience Puzzle.

Is it smaller than my hand?
Is it alive?
Does it move on its own?
Is it green?

Where Did It Live?

You go to a dry, dusty field with the paleontologist and a TV crew. The cameras are rolling. The host of the TV show points to a ditch. As you look closer, you see what looks like a fossil sticking out of the dirt. You ask your first question: Did the organism shown by the fossil live on land or in water? (Hint: It may look dry here now. But some deserts were once under the sea.)

A fossil found near a stream might have come from a plant.

The answer is: On land. That means that the fossil could have come from a plant. Some plants, such as the barrel cactus, grow on very dry land. Other plants, such as rice, grow in very wet areas.

Think about plants that live in wet places. How do you think they might look different from plants that live in dry places?

Many different creatures live inside saguaros.

Another possible consideration is that the fossil came from an animal. Animals live in many kinds of places. Bats live in caves. Prairie dogs live in underground tunnels. Some animals even live in plants! But other animals, such as fish, live in water.

You still don't know whether the fossil is of a plant or an animal. But now you know its life was on land.

These small plant leaves are plump and waxy, which helps them hold water.

How Did It Live?

Over many years, plants and animals **adapt**. That means they get better at living where they do. In a dry environment, plants have small leaves. Small leaves hold on to water really well. How do you think animals adapt to dry places?

You ask whether this fossil organism lived on dry or wet land. The host says this was a wet area. What might that tell you about how it lived?

What Did It Eat?

Whether they live in wet or dry environments, all living things need to eat. You ask whether the fossil organism made its own food. The host takes out a little brush. She is careful as she sweeps off the fossil. She is looking for clues that will help her answer your question.

To grow, plants need air, water, and **nutrients** from the soil. They also need sugars. Plant leaves use the energy in sunlight to make the sugars. This process is called **photosynthesis.**

The plant and trees in the picture make their own food with a little help from the Sun.

Leaves do more than just make sugars. They also help move the sugars to other parts of the plant. Sugars move through veins. Veins give structure to leaves.

Look at a leaf. You'll see one of two systems of veins. In a **dicot** system, veins branch off from one another. In a **monocot** system, veins run side by side.

Which vein system do you think offers more support for a leaf?

Moose are herbivores. They get all their large bodies need by eating many kinds of plants.

Plant Eater or Meat Eater?

The host looks at the fossil. She looks at you. She says, "No. The fossil organism did not make its own food." So now you want to know what it ate. If the answer is "plants," you will know the fossil is of an animal. Animals that eat only plants are called herbivores. Their teeth are strong and tough. Herbivores use their teeth mostly for grinding leaves, stems, and bark.

Can you name some animals that are herbivores?

What if you find out that the fossil organism ate animals? Chances are, it was an animal, too. Carnivores are animals that eat other animals. They often have sharp claws and teeth. They sometimes hunt in groups called packs. Lions and wolves are carnivores.

Carnivorous wolves hunt other animals for their meat.

The host keeps digging out the fossil. She has a better look now. She says the fossil organism ate animals and plants. So now you know it was an animal. Omnivores eat both animals and plants. Bears and raccoons are omnivores. Can you think of other types of omnivores?

How Did It Stay Safe?

Think about the animal your fossil came from. Its life may have been full of danger. Think about what you do to stay safe. You sleep under a roof. You wear a helmet when you ride your bike. How else do people avoid harm?

Some plants arm themselves with sharp thorns. Ouch! Other plants have poison in their leaves, flowers, or seeds. What do plants need protection from?

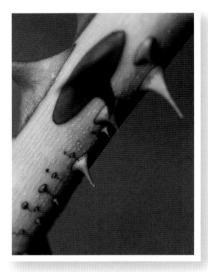

Roses are beautiful, but watch out for the thorns!

Did You Know?

Finding clues about past life can be hard work. Paleontologists work long hours. They often work in uncomfortable conditions and remote places. And they use a lot of tools. Shovels help them dig in dirt. Chisels and picks hammer through rock. Brushes move dirt off fossils. Magnifying lenses help paleontologists see little things. The scientists have to be very careful. Fossils can be fragile.

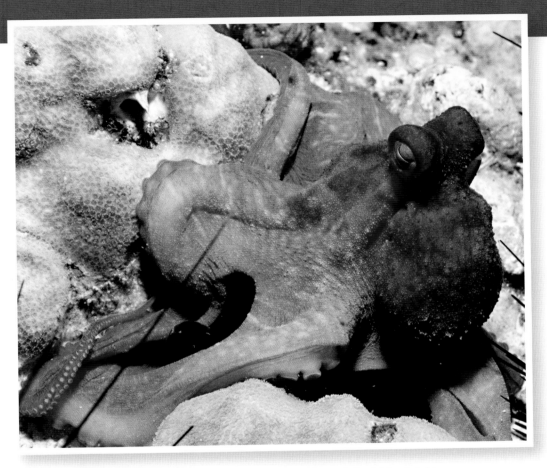

If you were to frighten this octopus, you'd get a face full of black ink!

Like plants, most animals face dangers. Animals have many ways to avoid getting eaten. Some, like porcupines, have sharp spines on their bodies for protection. Others use **camouflage** to hide. Some, like the octopus, shoot out clouds of ink so they can get away without being seen. A skunk sprays a nasty smell to drive other animals away.

Hard on the Outside

Some animals, such as lobsters and spiders, have an **exoskeleton.** That means that the hardest parts of their bodies are on the outside. For us, the hardest parts are our bones. They are on the inside. Other animals, such as turtles and armadillos, have hard outer shells for protection and bones on the inside. How else do animals protect themselves?

When frightened, an armadillo rolls itself into a ball so it is safe in its hard shell.

You ask the host how the fossil animal stayed safe. She answers by saying she has seen bones like this one before. Animals with this kind of bone had sharp spikey parts. Do you think you know yet what animal the fossil is from?

Connecting to History

Paleontologist Jack Horner

When famous director Steven Spielberg made *Jurassic Park* in 1993, he wanted to make sure he got the facts straight. He enlisted the help of a real paleontologist named Jack Horner. One of the characters in the movie was even based on him!

Jack Horner grew up in Montana. He discovered his first dinosaur bone when he was eight years old. When he was 13 he found a whole dinosaur skeleton. Horner never stopped studying dinosaurs. He went on to become one of the most well-known paleontologists in the United States. His work has challenged many old-fashioned ways of thinking about dinosaurs.

People can explore the way dinosaurs used to roam the Earth at the Calgary Zoo's Prehistoric Park in Alberta, Canada.

A Malasaura model shows how these creatures had bodies similar to large birds.

In 1978, on a dig in Montana, Jack Horner made an amazing discovery. He found the nesting grounds of a dinosaur species called Maiasaura. Hadrosaurs were duck-billed dinosaurs that lived around 70 million years ago. They traveled in large herds and grazed on low grasses. Until Horner's discovery, scientists believed dinosaurs abandoned their eggs and did not tend to their young. Horner found Maiasaura nests, eggs, and skeletons in "Egg Mountain," a huge Maiasaura breeding ground. The babies in the nest were twice as big as they would have been when they hatched. That meant their mother had been caring for them and keeping them safe.

Today, Horner is working on a dinosaur cloning project that would use chicken DNA to create a dinosaur. Jurassic Park may have been just a movie, but Jack Horner's work brings real dinosaurs to life in new ways.

Science at Work

Plant breeder

Plant breeders, also called geneticists, study ways to develop new plants and seeds. They combine the DNA of existing plants to make new ones. The new plants have traits that make them better in some way.

Plant breeders might develop a type of corn that is resistant to certain pests. They might breed tomatoes that don't need as much water so they can survive in drought seasons.

Corn has been bred over many generations to grow larger and have more corn kernels.

Plant breeders work in labs and greenhouses at large universities and at companies that package seeds. They advise farmers and others in the food industry about ways to get the most out of crops.

The New Generation

It's a fact of life. All kinds of living things, whether plant or animal, **reproduce.** That means they make **offspring**. Your animal, before it became your fossil, was formed through reproduction. It may have reproduced, too. In the process, **genes** pass from parents to offspring. This can happen in a variety of ways. Read on to learn about a few.

Lions and strawberry plants have something in common. They both make more organisms like themselves.

Plants have many ways of reproducing. Strawberry plants have runners. Runners are stems that grow along the ground. Runners can grow roots. New plants grow from the new roots. When plants grow this way, the offspring are exactly the same as the parents. Why do you think these special stems are called runners?

Seeds, Fruits, and Nuts

Another way of making offspring is with seeds. Most plants have seeds inside fruits or cones.

Some seeds have fluffy tails or wings. The wind carries them away. When they fall to the ground in a new place, they can grow into new plants.

Why do you think it might be good for a plant if animals eat its seeds? How might animals help seeds travel?

When you blow on a dandelion flower, you spread its seeds far away.

See the rows of spores underneath a fern leaf.

fern plant

Plants such as ferns usually use **spores** to reproduce. The spores typically lie on the bottom of their leaves. When spores fall to the ground, they are taking a step toward reproducing new ferns.

A fossilized March Fly was found near a fossilized leaf.

Sometimes, a single plant makes offspring from seeds. The young seedlings grow to look exactly like their parent plant. They have the same genes as their parent. Offspring made this way are called clones.

Other times, the genes of two parent plants mix together. This process makes new offspring, also from seeds. But in this case, the new seedlings grow to look different from their parents. Like this kind of plant, most animals reproduce by mixing genes from both parents.

Scientists often collect fossils from both plants and animals at the same time. Together, these remains can tell a lot about a time and place.

Hatching out of an egg is exhausting work!

Incredible Eggs

Now that you know about reproduction, you ask the host: How did the fossil organism create offspring when it was alive? She is still digging. Suddenly, she points to a broken object. "Eggs!" she says. Only animals make eggs.

Many birds lay their eggs in nests. Some frogs and salamanders attach their eggs to plants. Turtles bury their eggs in sand. Emperor penguins hold their eggs on their feet for months at a time! Even dinosaurs laid eggs. Babies grow inside eggs. After a while, they hatch.

It might help you to know a little more about eggs before you make your guess. Human mothers make eggs. So do dogs, horses, and other mammals. But mammals don't lay eggs. Instead, most mammal babies grow from eggs inside their mothers before birth. Human babies grow in their mothers for about nine months before birth.

A kangaroo baby is called a joey. This joey is learning about the world from the safety of his mother's pouch.

Kangaroo babies are born before they are fully developed. So, they crawl into their mothers' pouches. They grow more in there. Kangaroos are a specific type of mammal called marsupials. All marsupial females have pouches.

Solve the iScience Puzzle

The digging is done. The fossil is out of the rock. Look at the full fossil. Here's what you know.

- It lived on land in a wet place with lots of mud.
- It ate plants and animals.
- It had sharp spikes to protect itself.
- It probably laid eggs.

bone fossil

The fossil is clearly a bone. You know it came from an animal. But what kind of animal? You'll have to study the fossil more and compare it to other bones to find out for sure. But you have an idea. You know your fossil animal may have been a dinosaur! More than 65 million years ago, dinosaurs lived all over Earth. Some lived in wet places. Some ate plants and animals. All laid eggs.

You'll need to do more research and learn more facts to prove your idea, so keep digging!

Beyond the Puzzle

You have just learned some ways in which plants and animals differ. You have also learned that asking questions is a good way to solve mysteries in science. Now, try these activities:

- Find a lake, stream, or pond. Collect some water in small paper cups. Look at the water under a microscope or through a magnifying lens. Do you see creatures? Do you see plants? Can you find clues about past life? How can you tell what you are looking at? There is life in nearly every corner of Earth. You may be surprised at what you find.
- Imagine you are a paleontologist who lives 300 years in the future. Your school is now buried under the ground. Write an essay about a dig you do in the area. What do you find? What reasons do you give for what you find?

You can always ask questions of your own, too. What else do you want to know about plants and animals? What questions can you ask to find out what you want to know?

What can you tell about this dinosaur from its bones?

Glossary

adapt: change to fit a different condition.

camouflage: a covering or coloring that makes something look like its surroundings.

casts: molds made from holes left in rock by plant or animal remains.

dicot: a plant with leaf veins that branch out.

exoskeleton: a skeleton on the outside of an animal's body.

fossil: the remains of a plant or animal that lived a long time ago.

genes: material in a cell that controls inherited traits.

monocot: a plant with leaf veins that run next to each other.

nutrients: substances that plants and animals need to grow.

offspring: the young of living things.

organism: any one living thing.

paleontologist: a scientist who studies ancient life forms.

photosynthesis: the process in which plants use energy from sunlight to make food.

reproduce: to make offspring.

spores: special kinds of plant cells that can grow into a new plant.

Further Reading

Alkire, Jessie. 2018. *Discovering Fossils. Excavation Exploration.* Minneapolis, Minn.: ABDO Publishing Company.

Reed, Ellis M. 2019. *Dinosaurs Are Everywhere and Other Cool Jurrasic Facts.* Mind-Blowing Science Facts. North Mankato, Minn.: Bright Idea Books.

Sawyer, Ava. 2018. *Fossils. Rocks.* North Mankato, Minn.: Capstone Press.

Slepian, Curtis. 2019. *Digging Up Dinosaurs. Smithsonian Readers.* Huntington Beach, Calif.: Teacher Created Materials.

Additional Notes

The page references below provide answers to questions asked throughout the book. Questions whose answers will vary are not addressed.

Page 11: The fossil organism probably did not have to hold onto water very well.

Page 13: Caption question: A monocot vein system is stronger.

Page 16: People look both ways before crossing streets, they eat healthful foods, and avoid risks. Plants need protection from animals that might eat them.

Page 22: They run, or grow, along the top of the ground.

Index

camouflage, 17

carnivores, 15

fossils, 5, 6, 9, 10, 11, 12, 14, 15, 16, 18, 22, 25, 26, 28

genes, 22, 25

herbivores, 14

Horner, Jack, 19, 20

Maiasaura, 20

offspring, 22, 23, 25, 26

omnivores, 15

paleontologist, 6, 9, 16, 19, 29

photosynthesis, 12

plant breeders, 21

protection, 16, 17, 18, 28

reproduction, animals, 22, 25, 26, 27

reproduction, plants, 22, 23, 24, 25